Visit Susie and her Animal Friends at
KidsKyngdom.com

Visit Susie and her Animal Friends at
KidsKyngdom.com

To **Kelly**
Thank you for your love,
grace and compassion.

Visit Susie and her Animal Friends at
KidsKyngdom.com

Thank you for supporting Kids Kyngdom

Get your **FREE** audio book at <u>KidsKyngdom.com</u>

Send Dr. Susie a Creative Message on Instagram

I may pick You!

Instagram

Visit Susie and her Animal Friends at
KidsKyngdom.com

I'm here with **Bash**
Today we're visiting

Koalas in Australia

Visit Susie and her Animal Friends at
KidsKyngdom.com

Bash, did you know that I

named myself Doctor Susie?

If you could choose your own name

What would it be?

Visit Susie and her Animal Friends at
KidsKyngdom.com

Visit Susie and her Animal Friends at
KidsKyngdom.com

Robert?

That's a nice name

Visit Susie and her Animal Friends at
KidsKyngdom.com

Guess what name

a **baby koala** is born with

Visit Susie and her Animal Friends at

KidsKyngdom.com

Joe? That's right!

Baby Koalas are called Joeys

Visit Susie and her Animal Friends at
KidsKyngdom.com

Joeys are the size of a jelly bean when they are born

They have no hair, no ears, and can't see

They get to snuggle up all warm and safe

in their mommy's pouches

Visit Susie and her Animal Friends at
KidsKyngdom.com

Koalas have a great sense of **smell**

Visit Susie and her Animal Friends at
KidsKyngdom.com

They can find the tree they were born in just

by the way it **smells**

Hey Bash,
look at that Koala!

I wonder what it's **eating**

Visit Susie and her Animal Friends at
KidsKyngdom.com

Oh that's a Eucalyptus
(You Ka Lip Tus) **Gum tree!**

Let's call it a Gum Tree
for short - OK?

Koalas eat about **2 pounds** of gum leaves every day!

How much **water** do you think Koalas drink in 1 day?

That's Right!! They do not drink a lot of water

Visit Susie and her Animal Friends at
KidsKyngdom.com

Eucalyptus leaves
are like
watermelons!

Can you imagine drinking
a whole **watermelon**?

No, Bash stop

Eating eucalyptus can make us very sick
You would be **napping** all day

Visit Susie and her Animal Friends at
KidsKyngdom.com

Koalas mostly **nap** all day

They love to **sleep**

Not weep!
Sleep!

They **need** to sleep a lot

because gum leaves do not have a lot of **energy**

Koalas have **teeny tiny brains**

I wonder if it's because they have not eaten well over their **evolution**

Visit Susie and her Animal Friends at
KidsKyngdom.com

Remember our Gorilla

friends in Africa?

Visit Susie and her Animal Friends at
KidsKyngdom.com

They **eat a lot** of different foods

And they are so smart

Hey Bash
What kind of food do you not like?

Visit Susie and her Animal Friends at
KidsKyngdom.com

Crickets?

Ribbit

Visit Susie and her Animal Friends at
KidsKyngdom.com

Other animals do not eat Koalas because Koalas taste terrible

That's because Koalas eat **poisonous** gum leaves

So they are safe and comfy up in their tree

Visit Susie and her Animal Friends at

KidsKyngdom.com

They can see their habitat for many miles

Visit Susie and her Animal Friends at
KidsKyngdom.com

Hey Bash
Are you ready to cross the **Tasman** sea?

Visit the **United Kingdom?**

Where is it?

It's where we are going to visit **Swans**

Visit Susie and her Animal Friends at
KidsKyngdom.com

Sheep!
Swans!

Visit Susie and her Animal Friends at
KidsKyngdom.com

Do you know how many feathers a Swan has?

Search for Answers with Susie and Bash

Visit Susie and her Animal Friends at
KidsKyngdom.com

What is Sammie Kyng's Favorite Animal?

Sammie Kyng's favorite animal used to be a **Koala**.

Now it's a **Swan!**

If you have a favorite animal you would like Doctor Susie to visit, just send a request to Kidskyngdom.com or Instagram.

KidsKyngdom.Com

Instagram.com/Kids_Kyngdom

Visit Susie and her Animal Friends at
KidsKyngdom.com

ISBN: 978-1-959501-10-7

Published by: Kyngdom, LLC

Visit Susie and her Animal Friends at

KidsKyngdom.com

Did you love visiting **Koalas** with Susie and Bash?

Please Share Your Review

Visit Susie and her Animal Friends at
KidsKyngdom.com

Made in the USA
Monee, IL
05 November 2026

67377487R00026